Portage Public Library

Observing Bees
and Wasps

This book has been reviewed
for accuracy by
Walter L. Gojmerac
Professor of Entomology
University of Wisconsin—Madison.

Library of Congress Cataloging in Publication Data

Oda, Hidetomo.
  Observing bees and wasps.

  (Nature close-ups)
  Translation of: Hachi no kansatsu / text by Hidetomo
Oda; photographs by Hiroshi Ogawa.
  Summary: Discusses the life cycle, behavior patterns,
and habitats of various species of wasps and bees.
  1. Bees—Juvenile literature. 2. Wasps—Juvenile
literature. [1. Bees. 2. Wasps] I. Ogawa, Hiroshi,
ill. II. Title. III. Series.
QL565.2.O3513   1986   595.79   85-28195

ISBN 0-8172-2540-4 (lib. bdg.)
ISBN 0-8172-2565-X (softcover)

This edition first published in 1986 by Raintree Publishers.

Reprinted in 1989.

Text copyright © 1986 by Raintree Publishers, translated by Jun
Amano from *Observing Bees* copyright © 1983 by Hidetomo Oda.

Photographs copyright © 1983 by Hiroshi Ogawa.

World English translation rights for *Color Photo Books on Nature*
arranged with Kaisei-Sha through Japan Foreign-Rights Center.

All rights reserved. No part of this book may be reproduced or utilized
in any form or by any means, electronic or mechanical, including
photocopying, recording, or by any information storage and retrieval
system, without permission in writing from the Publisher. Inquiries
should be addressed to Raintree Publishers, 310 W. Wisconsin Avenue,
Milwaukee, Wisconsin 53203.

3 4 5 6 7 8 9 0   94 93 92 91 90 89

Observing Bees
and Wasps

Raintree Publishers
Milwaukee

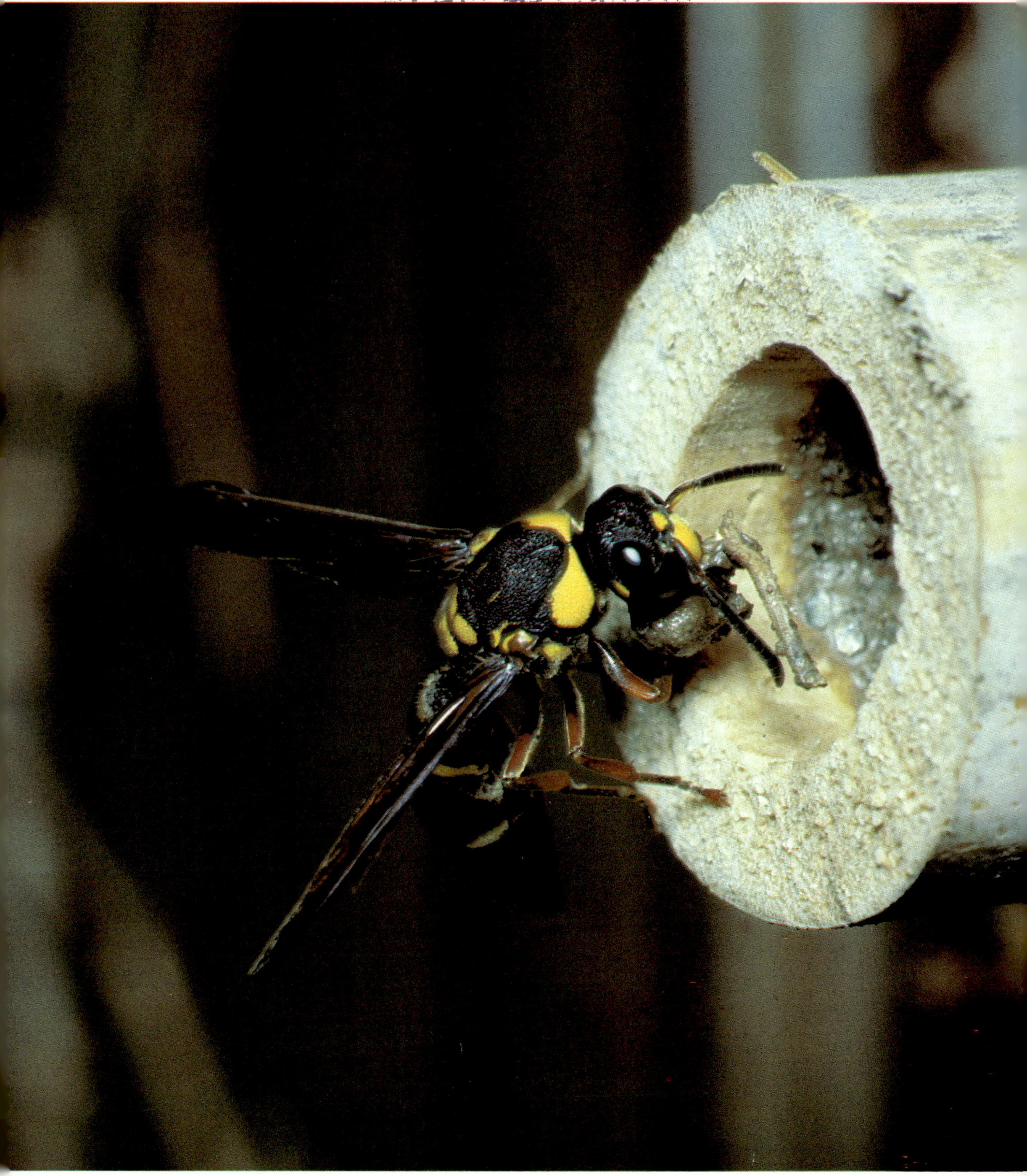

▶ **Hollow nest sites.**

Some kinds of wasps and bees look for hollow stems and twigs in which to build their nests. In some countries, nests are built in bamboo stems.

◀ **A mud wasp building a nest.**

This mud wasp will construct individual cells with mud walls inside this bamboo stem. The bamboo provides protection from enemies and bad weather.

Some insects live in large groups called colonies. These insects are called social insects because they live and work together. Bumblebees and honeybees are social insects. Worker bees build the nest, take care of the young, and protect the colony. The only task of the queen is to lay eggs.

But most kinds, or species, of bees—as well as most wasps and hornets—are solitary insects. That is, they live alone. A solitary queen bee or wasp works all alone to build a nest and take care of her young.

▶ **A bundle of bamboo stems.**

Different types of wasps and bees look for different nest sites. Generally, larger insects look for larger openings, and smaller insects look for smaller openings. In Japan, bees and wasps often build their nests in various sizes of bamboo stems.

6

◀ **A wasp carrying a caterpillar to the nest.**

The female wasp stings her victim to paralyze it. Then she carries it to the nest, holding it with her long legs and strong jaws.

▶ **Caterpillars stored in a nest.**

The wasp attaches its tiny egg to the ceiling of the nest with a silk thread. This ensures that the egg will not be disturbed if the caterpillars move around a bit.

Once the female wasp finds a suitable place to lay her eggs, she begins to build the nest and stock it with food. She instinctively knows what the right food for the young wasp will be, and she goes in search of it. Different species of bees and wasps prefer different types of food. Green caterpillars, moth larvae, and spiders are among the favorite foods of young wasps. Growing young bees like flower nectar and plant pollen.

▼ **A mud wasp catching a caterpillar.**

This species of mud wasp hunts green caterpillars for its young. The wasp frightens the caterpillar by nibbling its leafy nest. When the caterpillar comes out, the wasp seizes it.

▼ **A mud wasp stinging a caterpillar.**

The wasp stings the caterpillar several times in order to paralyze it. Then she carries it back to the nest.

▶ **A wasp carrying mud to her nest.**

The wasp uses her strong jaws to carry a ball of mud back to the nest site.

◀ **Two cells in a nest.**

The cells in this nest are separated by walls of mud. A tiny wasp egg hangs from the ceiling of each cell.

This mud wasp builds walls of mud to make individual cells inside a bamboo stem. She finds soil near the nest and moistens it with her saliva. The wasp uses her jaws to mix the soil and saliva to form mud. She then rolls the mud into a tiny ball and carries it back to the nest in her mouth. The wasp lays one egg in a cell. She then stocks each cell with caterpillars. When the eggs hatch, the young wasps will eat the caterpillars. Next, she builds a wall of mud to separate the cell from the next one. The wasp must carry many balls of soil to the nest before the nest is completed.

▼ **A mud wasp moistening the soil.**

This wasp wets the soil with her saliva and then rolls the mud into a ball.

▼ **A wasp rolling a ball of mud.**

The wasp uses her front legs and strong jaws to form a ball of mud.

9

◀ **This nest is almost sealed.**

The mud wasp gathers tiny balls of dirt to seal the nest. She holds the soil between her head and mid-section, or thorax, and spreads the dirt with her jaws.

▶ **A wasp sealing a nest (photos 1-3).**

The end wall of this nest may be about one-quarter of an inch thick when it is completed.

Once the mud wasp has finished making tiny, individual cells and stocking them with eggs and caterpillars, she begins to seal the end of the bamboo stem. She gathers many balls of wet dirt and takes them to the nest. Pressing down with her jaws and head, she spreads the mud to make a thick wall. Then she smears the mud with chewed-up tree fibers to make it stronger. When the walled-in end of the bamboo stem dries, it will be very hard. The thick wall helps keep the wasp eggs safe from enemies, and it helps to waterproof the nest.

▶ **Cells inside a bamboo stem.**

It takes the mother wasp two or three days to build five cells in a bamboo stem and seal the opening. Then the wasp looks for a place to build another nest.

◀ **An adult wasp sipping flower nectar.**

Although young wasps eat meat, adult wasps—like bees—prefer sweet flower nectar. Bees and wasps have mouths like straws, for sipping nectar.

1 2 3

11

◀ **Inside the egg, the larva develops.**

As the larva's body forms inside the egg, it becomes visible through the egg casing.

▶ **A larva emerges.**

The larva slowly wriggles its way, head-first, out of the egg. The larva is tiny, compared to the size of the caterpillars. This larva measures only about one-quarter of an inch long.

Like many insects, bees and wasps go through definite stages of development, called metamorphosis. The four phases of their life cycle include: egg, larva, pupa, and adult.

The length of time it takes for an egg to hatch varies, depending on the species of insect, and on the weather conditions. In about three days this wasp egg hatches. A tiny, wormlike larva wriggles head-first out of the egg. It has no eyes or legs, but it does have an enormous appetite. It immediately begins to feast on the caterpillars in the cell.

▶ **A tiny white wasp larva feeding on a caterpillar.**

◀ **Stored caterpillars in a wasp's nest.**

The paralyzed caterpillars pass waste material from their bodies (left photo). And sometimes they even shed their skin and enter the pupal stage of their development (right photo).

13

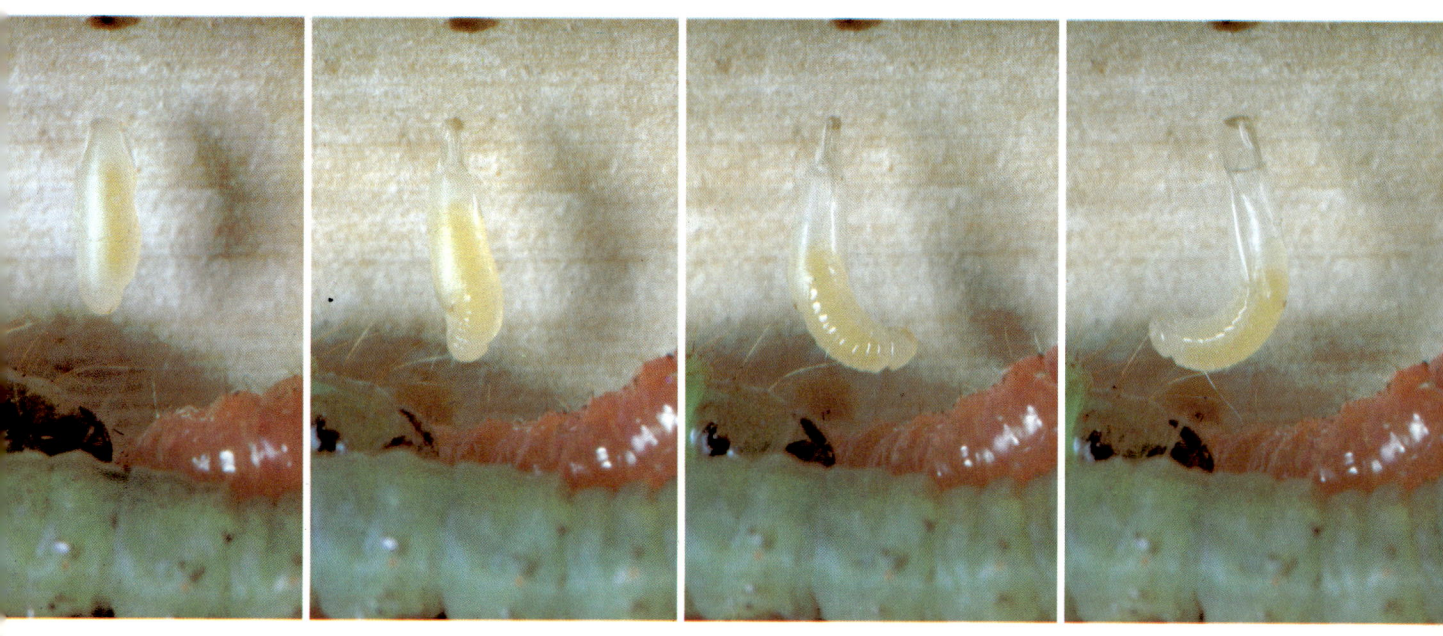

◀ **A larva surrounded by stored caterpillars.**

The larva doesn't need legs or eyes because it is surrounded by food. It doesn't have to go out and search for food, as many insect larvae do.

▶ **A larva which has molted three times.**

As it grows larger, the larva becomes hungrier. It eats most of the caterpillars in the cell after it has molted the third time.

The main activity of the wasp larva, or any insect larva, is to eat and grow larger. The wasp larva drinks the body fluids of the caterpillars with its mouth. Soon the larva becomes too big for its old skin. The wasp larva then molts, or sheds its skin. It molts three times during this stage of its development. As it grows larger, the larva eats almost constantly. It even eats the dried-up caterpillar skins. By the time the larva has eaten all the food in its cell, it has developed enough to enter the next stage in its life cycle.

▼ **A larva after its first molting.**

The wasp larva spends most of its time eating. The larva in this photo has molted once.

▼ **A larva after two moltings.**

The color of the wasp larva changes, depending on the color of the caterpillar it has just eaten.

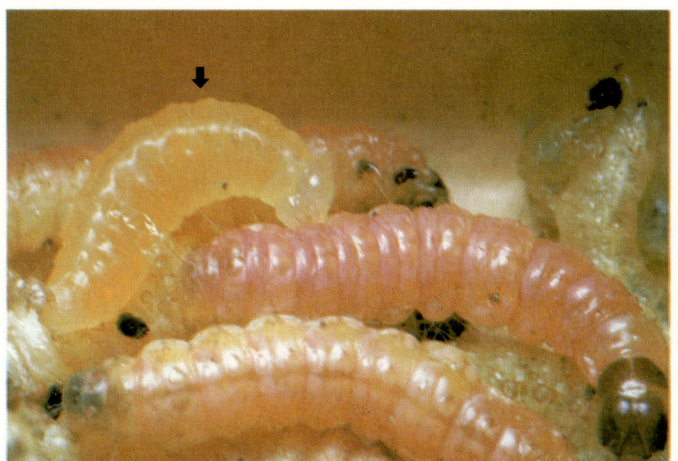

15

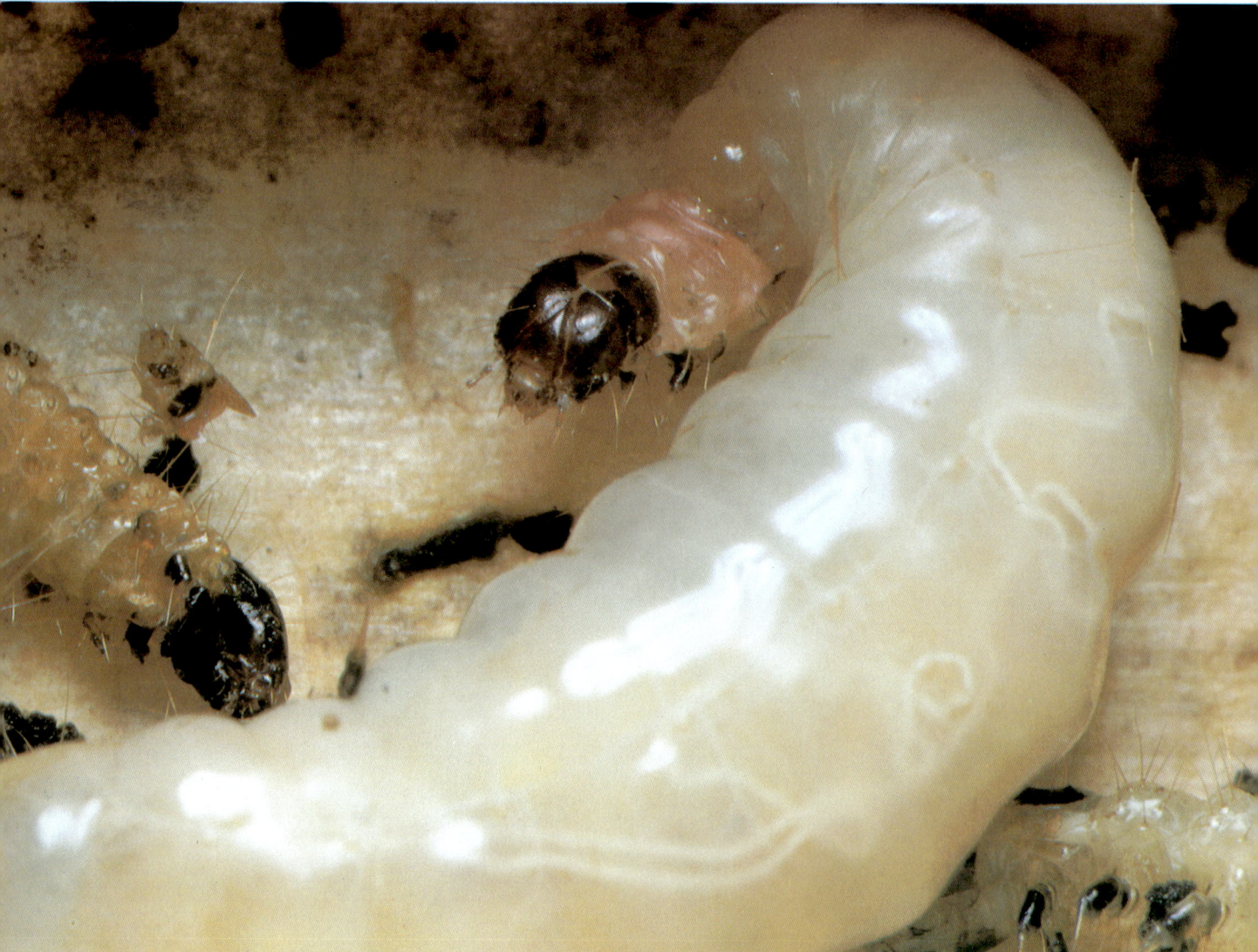

▲ **The pupa is pale in color at first.**

When the larva sheds its skin for the last time, it enters the pupal, or resting, stage.

▲ **An adult wasp shedding its pupal skin.**

When the body of the adult wasp has formed and darkened inside the pupal skin, the adult is ready to emerge. It sheds the pupal skin.

The wasp larva spins a cocoon for itself for the time it will spend in its resting, or pupal, stage. The cocoon is made of long, white, silky threads. They keep the pupa snug and warm. The pupal stage may last just a few weeks. But it may last for the entire winter, depending on whether the egg was laid in early summer or late fall.

During the pupal stage, mysterious changes are taking place. The long legs, powerful wings, and large eyes of the adult wasp are forming. When its body has fully matured and darkened in color, the wasp is ready to emerge and begin its life as an adult.

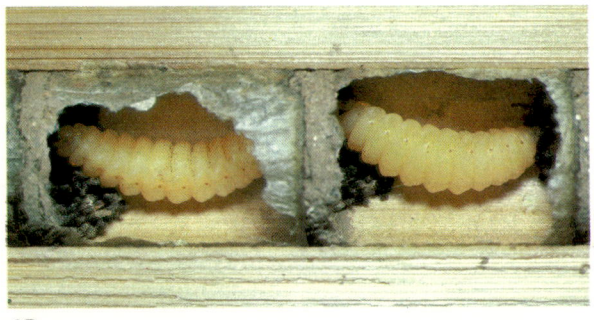

◀ **The larva spins a cocoon.**

The larva lines the walls of its cell with long, white strands of silk. Then it enters its resting stage, the third stage in its life cycle.

▼ **A newly emerged adult wasp.** Once it has shed its pupal skin, the adult wasp waits inside the cell until its wings have hardened. Then it chews its way out of the nest through the mud walls built by the female wasp.

◀ **A completed entranceway.**

The wasp carries many balls of dirt to build the entrance to the nest. Once the wet soil dries, it becomes as hard as clay.

▶ **A wasp building an entranceway.**

It takes more than a week for this species of mud wasp to complete a nest. There are several cells inside the nest.

This species of mud wasp also builds its nests in long, hollow spaces. But its nest has a hanging entranceway made of mud. Scientists believe this helps hide the nest from the wasp's enemies.

The wasp skillfully uses her strong jaws and long legs to form the mud cylinder at the end of this bamboo stem. She does not build divider walls inside the nest until her eggs have hatched and the larvae have molted for the first time. She continues to bring caterpillars to the nest until she has walled in a cell for each of her larvae.

▶ **A wasp carrying a caterpillar to the nest.**

This species of wasp continues to carry caterpillars to the nest after the wasp eggs hatch. The mother wasp doesn't divide the nest into cells until the larvae have molted once.

◀ **A wasp begins another nest.**

Sometimes a wasp uses the mud it has used in an earlier nest.

19

▲ **A wasp carrying a moth larva.**

This mother wasp carries a moth larva twice its size back to the nest.

▲ **A wasp sealing a nest.**

This type of wasp mixes small stones with wet soil to seal the nest.

Bees and wasps are very good at using the materials around them to build their nests. Moistened soil is often used to seal the opening to a nest. When it dries, it becomes as hard as clay. Sometimes, bits of dried leaves, tiny stones, and plant resin are mixed in with the soil. This "mortar" helps to make the seal even stronger. Inside, the wasp eggs and larvae are safe from enemies. And even during rainy seasons or cold weather, the young wasps stay snug and dry inside the weatherproof nest.

◀ **A wasp carrying a spider to its nest.**

This species of wasp hunts spiders to stock its nest. The long back section, or abdomen, extends out as the wasp flies.

▼ **A tiny wasp larva develops to the pupal stage.**

The mother wasp of this species lays her tiny egg on top of a single caterpillar in the nest. By the time the larva has consumed all the fluids from the two-inch-long caterpillar, it will be fully grown. It then spins a cocoon for itself and enters the pupal stage.

◀ **An adult wasp emerging from a nest.**

The wasp larva feels the roughness of the walls of the nest in order to determine the way out. It pupates facing in that direction. When the adult wasp emerges, it chews its way out of the nest.

▶ **An ichneumon wasp laying an egg in another wasp's nest.**

This ichneumon wasp holds her long ovipositor steady with her legs and drills through the wall of another wasp's nest. Then she lays her eggs inside.

Some insects, including species of wasps, do not build their own nests. Instead, they try to steal other nests in which to raise their young. Tachina flies, ichneumon wasps, and cuckoo wasps all try to steal nests of other bees and wasps.

The cuckoo wasp uses the long needle-like ovipositor at the end of its abdomen to drill through the walls of another wasp's nest. It lays its eggs on top of the pupating wasps. The eggs hatch into larvae and grow quickly as they feed on the young wasps in the nest.

▼ **Tachina fly pupae in a wasp's nest.**

The tachina fly lays its eggs in the wasp's nest while the female wasp is away. The fly larvae grow by feeding on the caterpillars the wasp has collected.

▼ **Chalcid wasp pupae in a wasp's nest.**

The chalcid wasp sneaks into another wasp's nest before it is sealed. She lays more than one hundred eggs on one caterpillar. When the eggs hatch, the larvae feast on the stored caterpillars.

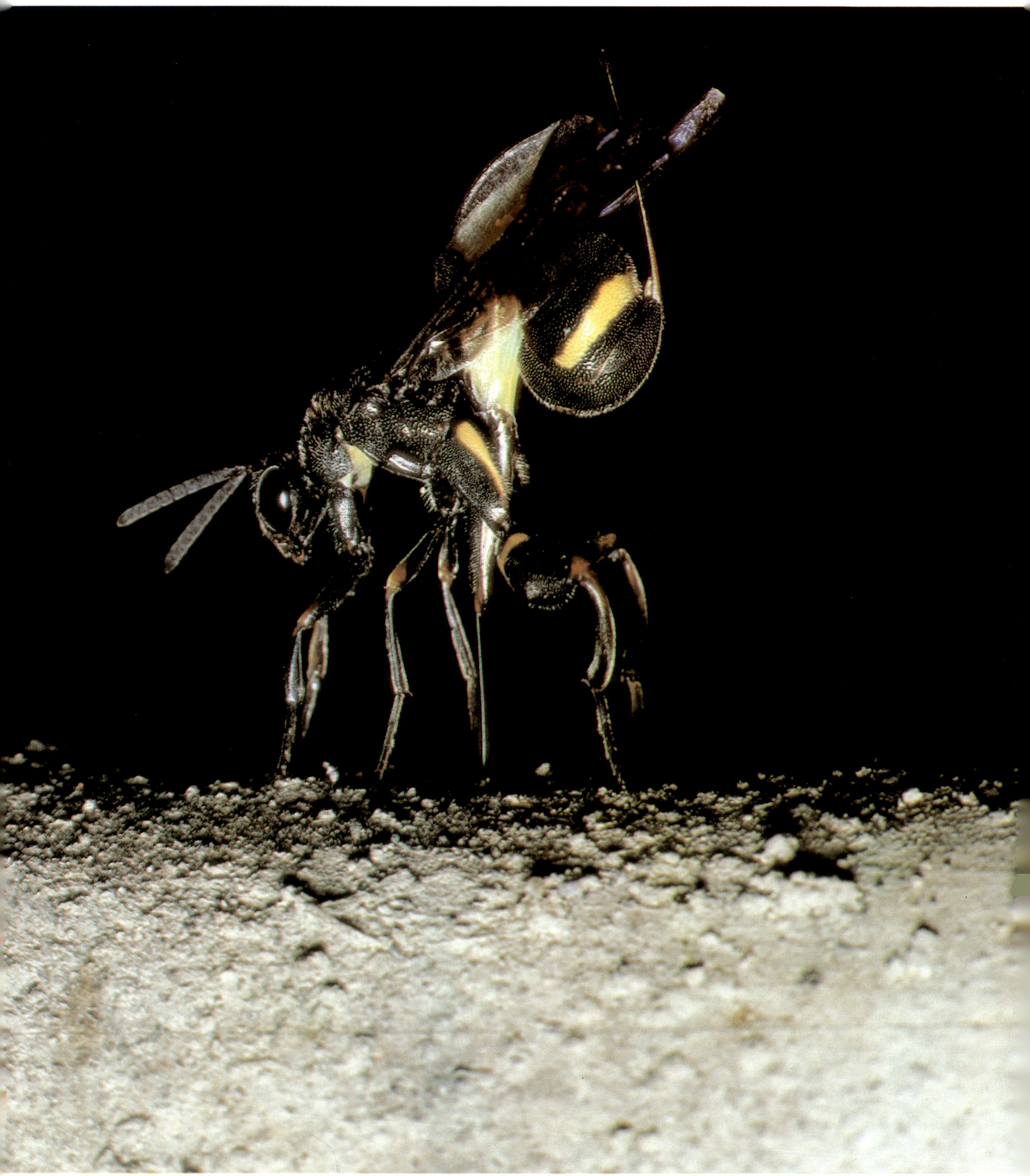

◀ **A bee carrying pollen to its nest.**

This species of bee carries pollen on the underside of its abdomen. Some species, including honeybees, roll the tiny grains of pollen into a ball. They carry it back to the nest in pollen baskets on their hind legs.

▶ **An egg in its nest.**

This species of bee uses soil and plant resin to build its nest in hollow spaces. When the cell is filled with a mixture of pollen and nectar, the mother bee lays her egg.

Wasps usually stock their nests with insects of some sort. But bees gather plant pollen and nectar for their young. Bees' mouths are well designed for drinking the sweet nectar of flowers. The nectar is stored in the bee's crop. As she moves about on the flowers, the bee is also busily collecting pollen. Tiny grains of it stick to the hairs on the underside of her abdomen. When she reaches the nest, the bee brings up, or regurgitates, the nectar from her crop. She mixes it with pollen and places some of the mixture in each cell of the nest. Then she lays her eggs.

▼ **A bee sipping nectar.**

Bees use their long, straw-like mouths for sipping sweet flower nectar.

▼ **A bee collecting pollen.**

Tiny pollen grains stick to the hairs on the underside of this bee's abdomen as it moves about on the flowers.

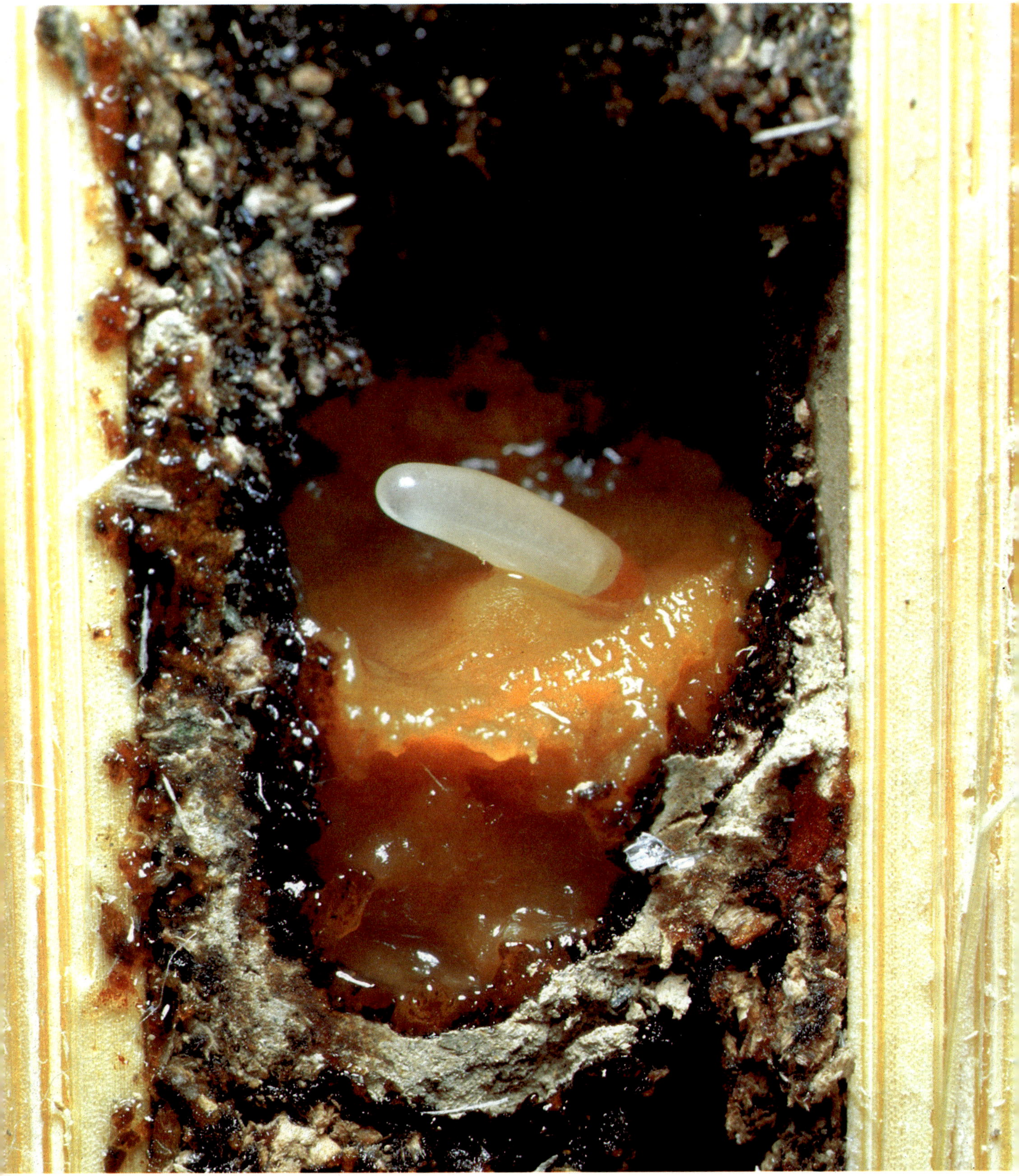

◀ **A leaf-cutting bee carrying a leaf.**

Leaf-cutting bees make their leafy nests in a variety of places, depending on the species. They may choose hollow plant stems, holes in tree bark, or bamboo stems.

▶ **A row of leafy nests.**

Leafy nests are placed side-by-side inside this bamboo stem. The moisture from the leaves prevents the stored pollen and nectar from drying out.

As their name suggests, leaf-cutting bees build their nests with leaves. The female bee begins to cut from the edge of the leaf, using her sharp jaws like a pair of scissors. When she has completely cut out a tiny round or oval piece, she rolls it up and flies with it to the nest. There she weaves pieces of leaves together to make tiny nests. Then the female bee gathers pollen and nectar and stores a mixture of it in each leaf nest. Finally, she lays her eggs.

▼ **A bee cutting a leaf.**

It takes only ten seconds for this tiny bee to completely cut out a piece of leaf.

▼ **A bee rolling a leaf.**

The bee rolls up the cut-out piece of leaf, tucks it between her legs, and flies back to the nest.

27

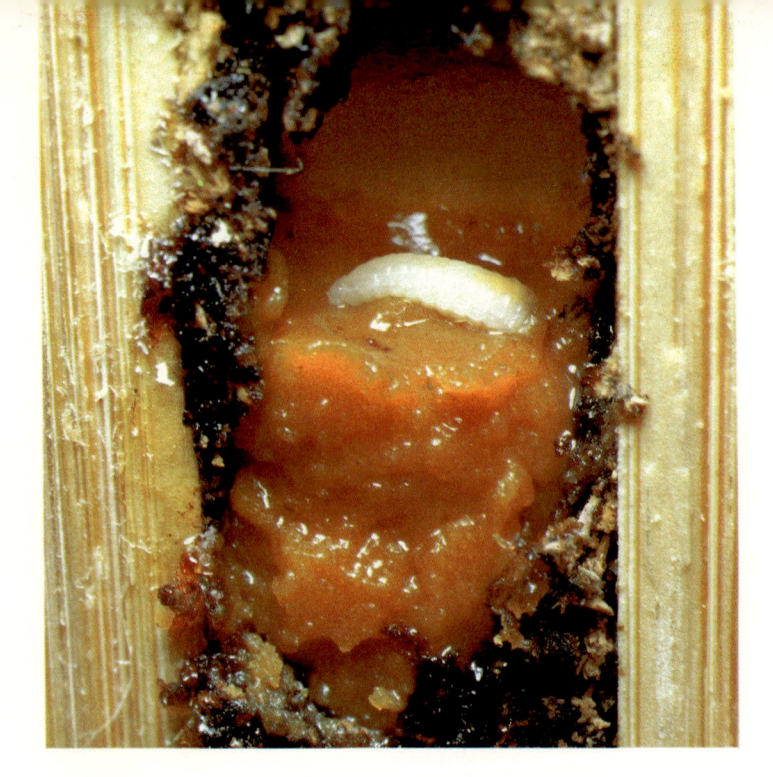

◀ **A larva which has just emerged.**

While wasp larvae get protein from feeding on other insect larvae, bee larvae get nutrition from pollen and nectar. This newly emerged larva eats the jellied pollen in its nest.

▶ **A larva grows bigger.**

The jellied pollen stays fresh until the larva eats it. The larva continues to eat, grow larger, and molt.

Once the tiny bee larvae emerge from the eggs, they grow quickly by feeding on the stored pollen and nectar inside the nest. Because plant pollen contains protein, it is very healthy for the young bees.

The nests pictured here are made of soil and plant resin. The resin helps make the nest airtight and also keeps the jellied pollen from drying out or becoming moldy. As the bee larvae eat and grow larger, they shed their old skin several times.

◀ **A bee collecting soil and resin.**

Bees use their mouths to gather nest materials. This bee gathers balls of soil (right), along with resin from the bark of trees (left).

▶ **Larvae growing inside the nests in a bamboo stem.**

29

◀ **Various kinds of nests.**

Just before winter, bees' and wasps' nests can be found everywhere. The nests here are made from a variety of materials. Some are sealed with mud, others are filled with wood shavings or dried leaves and grasses.

▶ **Larvae at various stages of growth.**

In the middle is a larva in a leaf-cutting bee's nest. To the right are nests filled with mature larvae who will spin cocoons soon. At this stage, the larva sheds the waste it has kept stored in its body. At the left are nests sealed with resin, stocked with pollen and nectar. Each bee larva has spun a cocoon.

Solitary wasps and bees continue their busy lives from summer through autumn. Few adults survive the winter. But inside the carefully constructed nests, eggs that were laid in the fall begin to hatch. Soon, the well-fed larvae grow to maturity. Then they spin cocoons and stay safe and warm inside the nest, no matter how cold the weather. In the spring when the temperature rises, the pupae begin to stir inside the cocoons. Soon, the adult bees and wasps will emerge from their nests and begin their busy lives.

▼ **A pupa.**

The body of the adult wasp, with its large eyes, long legs and wings, begins to form in this pupal stage.

▼ **An adult.**

This wasp's body has fully formed and darkened in color. Soon it will leave its cramped quarters inside the cocoon and will emerge as an adult.

GLOSSARY

abdomen—the back, or rear section, of an insect's body. (pp. 20, 22, 24)

cocoon—a protective casing spun by some insects during the pupal stage of their development. (pp. 16, 30)

colonies—large groups of insects that live together and depend on each other for survival. Ants and some bees live in colonies. (p. 5)

crop—a part of an insect's body used to store food. Both bees and ants have crops. (p. 24)

instinct—behavior with which an animal is born, rather than behavior that is learned. (p. 7)

metamorphosis—a process of development during which physical changes take place. Complete metamorphosis involves four stages: egg, larva, pupa, and adult. Incomplete metamorphosis occurs in three stages: egg, nymph, and adult. (p. 12)

molt—to shed the outer skin. (pp. 14, 18)

nectar—a sweet substance produced by flowers, eaten by insects such as butterflies and bees. (pp. 7, 10, 24)

pollen—the tiny grains that contain sperm cells which fertilize a plant's eggs. Pollen is a nutritious food for young, developing bees. (pp. 7, 24)

Observing bees and wasps /
J 595.79 O   31814850132736
Oda, Hidetomo.
PORTAGE PUBLIC LIB   10003

J595.79   Oda, Hidetomo
   O   Observing Bees And Wasps

                           Por. 10/95

Portage Public Library

PORTAGE